2. Fibre, Yarn and Fabric structures used in Industrial Textiles (35 pages)

2.1 Fibres in Industrial textiles
2.2 Yarn formation
2.3 Fabric Structures
   2.3.1 Woven Structures
   2.3.2 Knitted Structures
   2.3.3 Braided Structures
   2.3.4 Nonwoven Structures

2.1 Fibres in Industrial Textiles

In the last decade there was a remarkable growth of the production of industrial textiles, which definitively promoted this branch as one of the most potential and dynamical fields for development of the textile industry. The ranges of industrial textile products are very broad and it comprises various products such as industrial filters, hose & belts, ropes, acoustic materials, battery separators and composites. Today, we have access to a wide variety of fibres and yarns showing the appropriate characteristics required for producing high-tech textiles. The effects of different engineering and technological parameters on mechanical properties of high-modulus and high-strength polymer fibres and yarns are very important when designing industrial textiles. The fact that only small quantity of products (about 2–3 % by volume) utilize high-performance fibres, where the bulk of the products use conventional textile materials such as polyethylene (PE), polypropylene (PP), polyamide (PA), viscose, cotton, jute and glass. The technology of industrial textile is not always confined to the products themselves or to the production technology but it incorporates "know-how" for the application of these products for a broad range of end users. The development of industrial textile is closely associated to advancements in fibre production.

2.1.1 Natural Fibres

Natural fibres are obtained from plants, animals and geological processes. Plant fibres include cotton, flax, jute, bamboo, ramie, kapok, hemp and sisal among others. Flax is the oldest known fiber crop and is used for linen production. Jute fiber is the cheapest and strongest of all natural fiber and ranks second in production after cotton. It is used in traditional packaging fabrics, carpet backing, mats, bags, tarpaulins, ropes and twines etc. Manila is a bast fibre which finds application in ropes. Jute grids are used in drainage applications to separate contaminants from water. Cotton was used conventionally in hose manufacturing and low concentration filtration process. Decatising cloth made up of cotton blended fabric is used in decatising machine which is a part of mechanical finishing process of textiles. Bast fibres such as jute, linen and hemp are used in composites for short term applications owing to its biodegradability. It is typical that the natural fibres were heavy weight products with limited resistance to moisture, microbes, fungus and low flame resistance. These limitations reduce the possibility of use of natural fibres in industrial textiles.

2.1.2 Synthetic Fibres
a) Viscose rayon

Viscose rayon is the first commercial man made fibre used as reinforcing material for tires and other rubber products like safety belts, conveyor belts and hoses. The fibre has relatively high uniformity, tenacity (16 – 30 cN/tex) and modulus, especially if impregnated with rubber. The viscose fibre obtained by a special process of spinning has tenacity up to 40 cN/tex and elongation of 11–17 %. The tires designed for high quality roads still employ viscose fibre due to better thermal resistance. Majority of all rayon for polymer reinforcement is used as continuous filament, but there is still some use of spun staple rayon, where the main property required is bulk rather than strength. Viscose rayon is used as backing cloth in the coated abrasives owing to tensile strength and flexibility.

b) Polyamide

Polyamide fibre is characterized by high tenacity, elasticity, resistance to abrasion and moisture. Capability of energy resilience is a condition for an application in manufacturing climbing ropes and linen for parachutes and sail fabrics. The typical application of polyamide is for reinforcing tires for use at low quality roads and of road vehicles. Nylon is used as facing fabric for the transmission belts due to its better abrasion property. Polyamide fibres, due to its poor acid resistance, its usage is limited in industrial filtration. Polyamides are used as carcass material in conveyor as well as transmission belts due to its better adhesion property. The better abrasion resistance of polyamides finds its application in the manufacturing of industrial brushes. Bolting fabrics made up of nylon is used as screen in screen printing operation. The Nylon 6 yarn is woven into a fabric which is cut to required size for making computer printer ribbons. The reason attributed for the selection of nylon 6 is tensile strength, capillary action, scratch resistance and heat resistance.

c) Polyester

Production of polyester made possible to get fibre for technical application at a lower price compared to polyamide and viscose fibre. Polyester is used as reinforcement material in tire cord, hose and conveyor belts because of its superior mechanical properties. Polyester has good resistance towards acid and moisture which find its application in liquid filtration process of highly acidic in nature. Due to its excellent moisture resistance, it is used as reinforcement fabric in water hoses. Polyester cords are widely used as the carcass material in the transmission belts. The lower modulus, elasticity and better shock absorption property of polyester enables the belt to rotate smoothly over small diameter pulleys. Polyester is used as backing cloth in coated abrasives. Polyester fabrics are used as paper making fabrics due to its good drainability, abrasion resistance and moisture resistance.

d) Polyolefin fibres

Polyethylene and Polypropylene are the polyolefin fibres contribute significantly to the industrial textiles. Advantages of polyolefin fibres are low price, low specific gravity, good abrasion resistance, and low moisture content. These properties have determined their usage in a range of technical applications such as ropes, filter fabrics, nets, etc...Polyethylene is used in battery separators due to its excellent chemical resistance.

e) High performance fibres

The development of carbon fibres and aramid fibres in the 60's triggered many developments of high performance fibres and yarns. Today access to a wide variety of fibres and yarns showing the appropriate characteristics required for producing high-tech textiles. These include high modulus/high tenacity, heat resistance and stability to chemicals even at elevated temperatures. Aramid is used as reinforcement material in the specialty conveyor belt where high strengths and modulus are required. The typical properties of Aramid fibre are low density, high strength, good impact resistance, good abrasion resistance, good chemical resistance, good resistance to thermal degradation and compressive strength similar to E-glass fibres. The characteristic property of aramid fibre is a high melting temperature of 370 °C (compared to 248 °C at conventional polyamide). Due to such a property the use of aramid fibre is extended to high temperature applications. Polyethylene processed by extended highly oriented chain structure has got much higher strength. The extension of polymer chains and high longitudinal orientation is a precondition for accomplishing high mechanical properties. The result of this treatment is production of high-performance polyethylene fibre (HPPE), of so far the highest strength of 400 cN/tex, i.e. two times higher than aramid fibre.

The fibre is later commercialized under several versions of which the most known are Dynema and Spectra. The advantages of HPPE are low specific gravity (0.396 g/cm^3), almost about the half less than a high modulus carbon fibre and about one third less than the aramid fibre. The fibre has low melting temperature (~150 °C), which restricts the possibility for the high temperature application. Carbon fibre can be manufactured from several precursors, of which rayon and acrylic are the most usually employed. Carbon reinforced composites find application in civil aviation, special sport and industrial goods, such as turbine parts for generators and reinforced fuel tanks. Glass fibre is at great extent accepted in the production of high-performance composite materials, including protective materials, various filters, protective clothing and packing. Nonwoven glass mats are used as battery separators due to its superior resistance to chemical, oxidation and contaminants. Dimensional stability and high strength to weight ratio of glass fibre is also a reason for its use in battery separators and many composite applications. Glass fibre prepregs are used in printed circuit board (PCB) owing to its uniform dielectric constant, lower dissipation factor and dimensional stability.

The development of technical textile is closely associated to advancements in fibre production. The creation of polyamide fibre (Carothers, 1930), gave a direction for development of polymer technology, followed by the invention of polyester, polyethylene, polypropylene and carbon fibres. In recent time, high-performance fibres as Aramid, UHMW-polyamide, HP-polyethylene, that had an extraordinary significant influence for development of industrial textile were obtained. It is to expect that the knowledge gained so far from manufacturing high-performance fibre would be of benefit in realizing the predetermined goal of processing technical fibre of fantastic tenacity of 900 cN/tex (100 grams per denier). This would mean that a great number of products the metals and other traditional constructive materials would be replaced. The typical properties of some high-performance fibres are depicted in Table 1.

Table 1. Mechanical characteristics of high=performance fibres

Fibre type	Tensile strength MPa	Tensile modulus GPa	Density g/cm³	Failure strain %
Aramid	3450-3620	112-179	1.44-1.47	1.9-6.2
E glass	3448	72.4	2.54	3.5
Vectran HS	3210	135	1.41	2.3
Carbon	2900-4800	230-390	1.74-1.81	0.7-1.8
HPPE	3090	172	0.97	1.8

Numerous multifunctional fibres are nowadays available on the market, offering a diversity of improved functional properties. In addition to thermally adaptable fibres (e.g. hollow high-loft PET fibres; PP/PET blend fibres; hollow fibres containing water soluble phase change materials, etc,), a new generation of fibres based on a multiproperty holistic concept are developed for the use in automotive interiors, battery warmer, outdoor architectural structures, protective clothing for bullet proof vests, geotextile, agricultural etc.

f) Specialty Fibres

Bicomponent technologies have made significant advancements since their introduction in the mid-20th century. Fibres with different cross sections to enlarge the fibre surface enhancing the performance are increasingly being used which are shown in Figure 1. Fibres which show a profiled cross section possess a wider specific surface which makes the separation of particle smaller than 5micron more effective. Factors influencing these characteristics are the cross section, shape and micro fibrillation of such materials. Both finest and micro fibres are, because of their filter surface, preferably arranged and used to enlarge the effective filter surface.

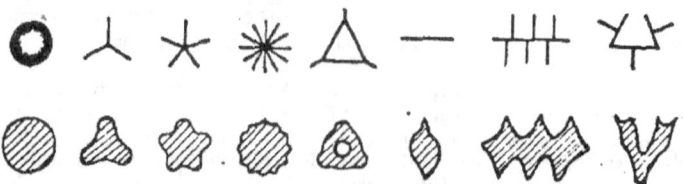

Figure 1 Polyamide fibres of various cross section

2.2 Yarn Formation

Yarn is continuous strand which is made up of filaments or fibres. It is used to make textiles of different kinds. Yarn formation methods were originally developed for spinning of natural fibres including cotton, linen, wool and silk. Since the overall physical characteristics of the fibres and processing factors needed differed from fibre to fibre, separate processing systems were developed. As synthetic fibres were introduced, synthetic spinning systems for texturised and untexturized cut staple were developed as modifications of existing staple systems, whereas spinning systems for texturised and untexturized filament were developed separately. Yarn has different form such as staple fibres, monofilament, multifilament, tow and textured yarn.

Staple fibre yarns

Staple fibre yarns can be manufactured in short staple and long staple spinning systems. In natural fibre spinning system, fibre bale is opened, cleaned and converted to sliver form. The sliver is then drafted and twisted into roving and subsequently into yarn form. The object of spinning and of the processes that precede it is to transform the single fibres into a cohesive and workable continuous-length yarn. Yarn making from staple fibers involves picking (opening, sorting, cleaning, blending), carding and combing (separating and aligning), drawing (re-blending), drafting (drawing into a long strand) and spinning (further drawing and twisting) Staple yarns, made from shorter fibers require more twist to provide a sufficiently strong yarn; filaments have less need to be tightly twisted. For any fiber, yarns with a smaller amount of twist produce fabrics with a softer surface; yarns with considerable twist, hard-twisted yarns, provide a fabric with a more wear resistant surface and better resistance to wrinkles and dirt, but with a greater tendency to shrinkage. Hosiery and crepe fabrics are made from hard twisted yarns. Short-staple spinning is the logical development of the cotton spinning of history, but the range of fibers has increased dramatically in this century. Short staple spun yarns can be manufactured by several spinning techniques such as ring, rotor, air jet, friction and twist less spinning. The main advantage of staple fibre spinning system is the possibility of altering different fibre blend ratio to achieve the desirable properties in the yarn.

Staple fibre yarn structure is predominantly used in filtration process which needs high dust retention capacity. Manufactured fibers used in textile manufacture come from both natural and man-made sources. Natural sources are either organic or inorganic. Organic materials include those from plant cellulose or rubber and those from manufactured polymers. Those from polymers, derived primarily from petroleum, coal and natural gas, include polyesters, acrylics, nylon, polyethylene, polypropylene, polyvinylchloride, polyurethane and synthetic rubbers. Synthetic fibers made from cellulose include rayon, acetate and triacetate. Inorganic fiber materials include metal and glass. Man-made staple fibers are made from tow, which is extruded in the same basic way as with filament yarns; however, the number of filaments involved is vastly larger. Filament tow is cut into staple fibres according to the end use requirement. Synthetic staple fibres usually arrive at the mill in compacted bales containing about 500 lb of fibre. Synthetic staple fibres are blended with other fibre in blow room or draw frame and then subsequently processed in speed frame and ring frame machines. Spinning process involves drafting, twisting and winding of the yarn into the cop.

Filament Production
The term "spinning" is also used to refer to the extrusion process of making synthetic fibers by forcing a liquid or semi-liquid polymer through small holes in an extrusion die, called a spinneret, and then cooling, drying or coagulating the resulting filaments. The fibers are then drawn to a greater length to align the molecules to increase their strength. The monofilament fibers may be used directly as-is, or may be cut into shorter lengths, crimped into irregular shapes and spun with methods similar to those used with natural fibers. Extruded filament yarn manufacture is a short, mechanical process involving only one or two steps. The term spinning here defines the extrusion process through spinnerets of fluid polymer masses which are able to solidify in a continuous flow. The polymer processing from the solid to the fluid state can take place with two methods:

- **a) Melting**: this method can be applied on thermoplastic polymers which show stable performances at the processing temperatures.
- **b) Solution**: the polymer is dissolved in variable concentrations of solvent according to the kind of polymer. Solvent evaporation is carried out by either coagulation or evaporation.

The three main stages for melt spinning polymers are preparation of the melt, extrusion and winding. A molten polymer is forced through a spinneret orifice at a given temperature, pressure and rate. The flow is collected at a different velocity at the site of 'take-up.' The distance between the spinneret and take-up is variable. Once the polymer reaches the take-up area, the process of initial fiber formation through solidification and cooling is finished. Solution spinning is classified into wet spinning and dry spinning. In the wet-spinning process, dissolved polymer solution is pumped through a spinneret that is submerged in a coagulating bath. The bath contains the solvent and water. As the polymer solution passes through the coagulant or non solvent, a phase change occurs whereby the solvent diffuses out and the non solvent diffuses in. The newly formed fibers emerge in a gel form from the bath where they are later subjected to a series of after treatments. In contrast to wet spinning where non solvents diffuse into and solvents diffuse out of the dope, formation of fibers by dry spinning occurs from solvent evaporation. Evaporation is encouraged through a hot inert gas as the gel passes out of the spinnerets.

Textured yarn production
Texturisation is the process of imparting crimp to the manmade fibre. In this way the yarn contains the many air pockets needed to produce insulation properties, permeability, and softness. Furthermore, the yarn now occupies a greater volume; the greater the bulk, the better the cover. Also the yarn becomes more extensible and this, too, is an added attraction. It is possible to get various combinations of stretch and bulk. Texturisation process is done with several techniques such as false twist, air jet, knit-de-knit, gear, stuffer box and draw texturisation.

Doubled yarn production
For sewing threads, as well as certain specialty and industrial yarns, it is necessary to ply (i.e. to double or fold) the yarns to give them a smoother and less hairy character. Doubling improves the evenness; plying balances torque if carried out correctly and binds some of the hairs on the component yarns. The traditional methods include assembly winding to place the single yarns parallel to each other as a closely spaced pair (or group) of yarns on an intermediate package. The new package is then used as a feed for a twisting machine and the output is a plied yarn. However, the cost of assembly winding approaches 25% of the total winding costs and the system is prone to problems.

2.2 Fabric Structures

2.2.1 Fabric Formation

Fabric formation technology can be broadly classified into weaving, knitting and nonwoven.

Weaving
The conversion of yarn into woven fabric is accomplished by interlacing warp and the weft on a weaving machine or loom. In a weaving machine, the warp yarns are passed from a warp beam to the fabric beam. During this process, each warp yarn is led through an eye on a heald attached to a harness. The harness lifts some of the warp yarns and depresses the remainder to form a gap between them known as a "shed" through which the weft is inserted. This operation is known as shedding and the insertion of weft is called "picking". The sley beats up the weft to the edge of the woven fabric (Figure 2).

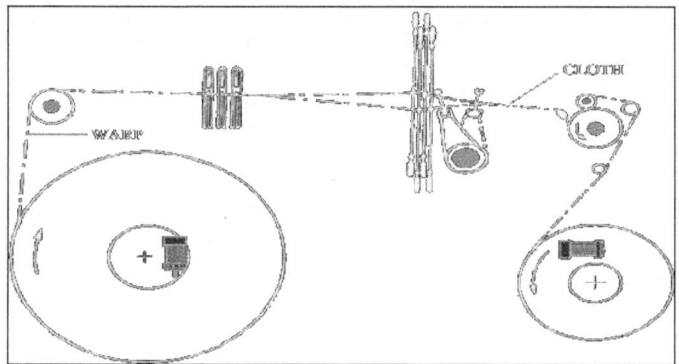

Figure 2. Weaving mechanism of a plain loom

Weaving machines are classified according to the type of weft insertion system which at present includes shuttles, rapiers, projectiles, air jets and water jets. Woven fabrics find application in high velocity filtration where mechanical properties of filter fabric are crucial in deciding life of filter fabric. Woven fabrics are used as reinforcements in hoses, conveyor & transmission belts and composite materials owing to its superior mechanical properties.

Woven Structures

Woven structures have the greatest history of application in textile manufacturing. Woven fabrics are made on looms in a variety of weights, weaves and widths. Woven structures are bidirectional, providing good strength in the direction of warp yarn. The threads that run along the length of the fabric are called warp or ends, while the threads that run along the width of the fabric from selvedge to selvedge are referred as weft or picks. Woven fabrics are produced by the interlacing of warp (0°) fibres and weft (90°) fibres in a regular pattern or weave style. The fabric's integrity is maintained by the mechanical interlocking of the fibres. Mechanical properties of woven fabrics, which are especially important for industrial textile, depend on type of raw materials, type and count of warp and weft yarns, yarn density and the type of weave structure. Drape (the ability of a fabric to conform to a complex surface), surface smoothness and stability of a fabric are controlled primarily by the weave style. However, the tensile strength of woven fabrics is compromised to some degree because fibres are crimped as they pass over and under one another during the weaving process. The crimp influences the fiber volume fraction, fabric thickness and fabric mechanical properties. Fibre volume fraction and fabric thickness are the important process parameter influencing the properties of textile composites. The cloths are lighter in weight, typically from 6 to 10 ounces per square yard and require about 40 to 50 plies to achieve a one inch thickness. Impact resistance is enhanced because the fibres are continuously woven. Fabric area density and cover factor influences strength, thickness, stiffness, stability, porosity, filtering quality and abrasion resistance of fabrics. The woven structures can be broadly classified as:

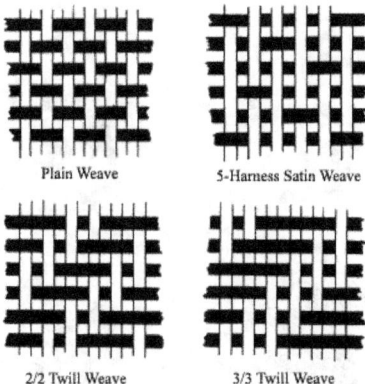

FIGURE 3 TYPICAL TRADITIONAL WOVEN FABRIC CONSTRUCTIONS

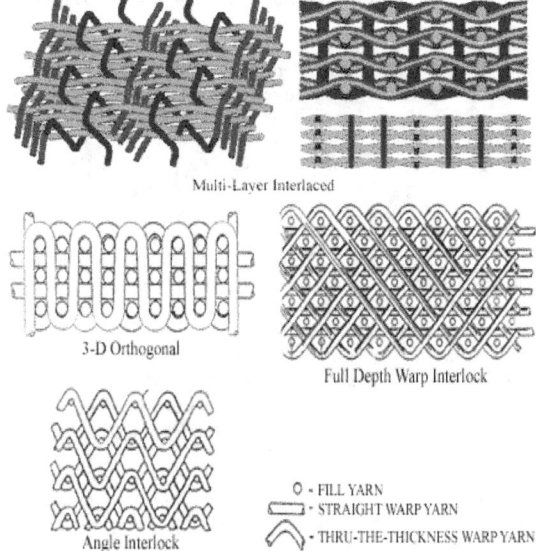

FIGURE 4 CONSTRUCTIONS OF MULTIDIRECTIONAL WOVEN FABRICS

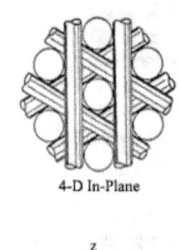
4-D In-Plane

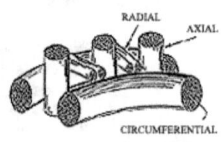
3-D Cylindrical

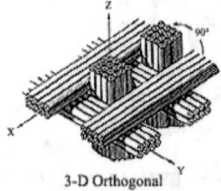
3-D Orthogonal

4-D Pyramidal

FIGURE 5 ADVANCED WOVEN STRUCTURES FOR INDUSTRAIL TEXTILES

a) Plain Weave

The structure where warp yarns alternatively lift and go over across one weft yarn and vice versa is the simplest woven structure called plain weave. However, it is the most difficult of the weaves to drape, and the high level of fibre crimp imparts relatively low mechanical properties compared with the other weave styles. With large fibres (high tex) this weave style gives excessive crimp and therefore it tends not to be used for very heavy fabrics.

b) Twill Weave

Twill is a weave that produces diagonal lines on the face of a fabric. One or more warp yarns alternately weave over and under two or more weft yarns in a regular repeated manner. The direction of the diagonal lines viewed along the warp direction can be from upwards to the right or to the left making Z or S twill. This produces the visual effect of a straight or broken diagonal 'rib' to the fabric. Compared to plain weave of the same cloth parameters, twills have longer floats, fewer intersections and a more open construction. Superior wet out and drape is better in the twill weave over the plain weave.

c) Satin Weave

A weave where binding places arranged to produce a smooth fabric surface free from twill lines is called satin. The 'harness' number used in the designation (typically 4, 5 and 8) is the total number of yarns crossed and passed under, before the yarn repeats the pattern. The 5 ends satin is most frequently used for technical applications for providing firm fabric although having moderate cover factor. Satin weaves are very flat, have good wet out and a high degree of drape. The low crimp gives good mechanical properties. Satin weaves allow fibres to be woven in the closest proximity and can produce fabrics with a close 'tight' weave. However, the style's low stability and asymmetry needs to be considered.

d) **Leno weave**

A form of plain weave in which adjacent warp yarns are twisted around consecutive weft yarns to form a spiral pair, effectively 'locking' each weft in place; thus securing a firm hold on the filling yarn and preventing them from slipping out of position. It is also called the gauze weave. Leno weave improves the stability in 'open' fabrics which have a low fibre count (Figure 6).

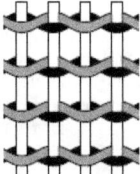

Figure 6 Leno weave

2.2.2 Non-crimp Fabrics

In non-crimp fabrics, yarns are placed parallel to each other and then stitched together using polyester thread. Warp unidirectional fabric is used when fibres are needed in one direction only, for example, in stiffness-critical applications such as water ski applications where the fabric is laid along the length of the ski to improve resistance to bending (Figure 7). Non-crimp fabrics offer greater flexibility compared to woven fabrics. Non-crimp fabrics offer greater strength because fibres remain straight; whereas in woven fabrics, fibres bend over each other. Non-crimp fabrics are available in a thick layer and thus an entire laminate could be achieved in a single-layer fabric.

Figure 7 Non-crimp fabric

2.2.3 Knitted Fabrics

Knitting is the second most frequently used method, after weaving, that turns yarns or threads into fabrics. It is a versatile technique that can make fabrics having various properties such as wrinkle-resistance, stretchability, better fit, particularly demanded due to the rising popularity of sportswear and casual wears. As of present day, knitted fabrics are used widely for making hosiery, underwear, sweaters, slacks, suits and coats apart from rugs and other home furnishings. A knitted fabric may be made with a single yarn which is formed into interlocking loops with the help of hooked needles. According to the purpose of the fabric, the loops may be loosely or closely constructed. Knitted fabrics are textile structures assembled from basic construction units called loops. There exist two basic technologies for manufacturing knitted structures: weft and warp knitted technology. Knitted fabrics are easy to handle and can be cut without falling apart. A knitted reinforcement is constructed using a combination of unidirectional reinforcements that are stitched together with a nonstructural synthetic such as polyester. A layer of mat may also be incorporated into the construction. The process provides the advantage of having the reinforcing fibre lying flat versus the crimped orientation of woven structure. Knitted fibres are most commonly used to reinforce flat sections or sheets of composites, but complex 3-dimensional performs have been created by using prepreg yarn.

Weft knitted fabrics

The feature of the weft knitted fabric is that the loops of one row of fabric are formed from the same yarn. A horizontal row of loops in a knitted fabric is called a course and vertical row of loops is called a wale. The stitch density is the number of stitches per unit area in the knitted fabric. The stitch length is the length of a yarn in a knitted loop and is an important factor that determines the properties of the weft knitted fabric. The cover factor is a number that indicates the extent to which the area of a knitted fabric is covered by the yarn. The higher cover factor indicates a more tight structure and vice-versa. The fabric area density is a measure of the mass per unit area of the fabric. In weft knitted fabrics the loops are formed successfully along the fabric width. The yarn is introduced more or less under the right angle regarding the direction of the fabric formation. The feature of the weft knitted fabric is that the neighboring loops of one course are created of the same yarn (Figure 8). The simplest weft knit structure produced by the needles of one needle bed machine is called plain knit or jersey knit. The plain knit has different appearance of both sides of the fabric. A structure produced by the needles of both needle beds is called rib structure or double jersey having the same appearance on both sides of the fabric. Weft knitted fabric can be produced on a number of different types of knitting machines. Circular or flat bar machines using a latch needle can produce both fabrics and knitting garments. Straight bar or circular machines using a bearded needle can produce shaped knitwear. Many machines can produce a double fabric structure with differing knitted structure on each fabric face. Spacer yarns can be inserted between the front and back fabric, thus creating a complex three layer structure. The properties of each layer are determined by the fibre, yarn properties and structure of that layer. These structures can be tailored for specific applications and are useful in the protective textile field (Figure 9).

In the more common weft knitting, the wales are perpendicular to the course of the yarn. In warp knitting, the wales and courses run roughly parallel. In weft knitting, the entire fabric may be produced from a single yarn, by adding stitches to each wale in turn, moving across the fabric as in a raster scan. By contrast, in warp knitting, one yarn is required for every wale. Warp-knitted fabrics such as tricot and milanese are resistant to runs, and are commonly used in lingerie.

Knit stitch

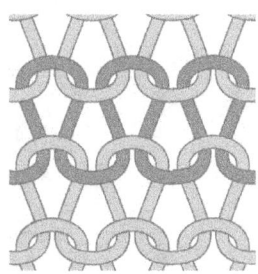
Purl stitch

Missed stitch

Tuck stitch

Figure 8 Weft knitted fabric structures

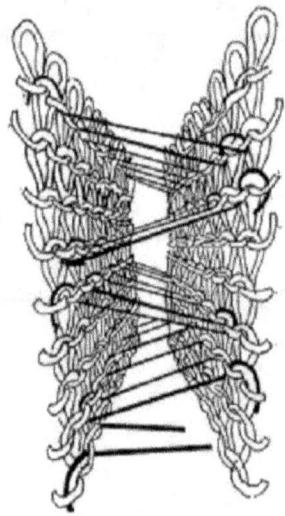

Figure 9 Double layered weft knitted fabric construction

Warp knitted fabrics

In warp knitted technology every loop in the fabric structure is formed from a separate yarn called warp which mainly introduced in the longitudinal fabric direction. The most characteristic feature of the warp knitted fabric is that neighboring loops of one course are not created from the same yarn. To accomplish the warp knitted structure every needle along the width of the fabric must receive yarn from the individual guide. The function of the guide is to lead and wrap the warp yarn around the knitting needle during the knitting process. The loop structure in the warp knitted and weft knitted structure is similar in appearance (Figure 10). The warp knitted structure is very flexible and regarding construction it can be elastic or inelastic. The mechanical properties are in many cases similar to those of woven structures. The best description of warp knitted fabrics is that they combine the technological, production and commercial advantages of woven and weft knitted fabrics. Warp knitted fabrics can be produced on a number of different types of knitting machine. Raschel machines using a latch or compound needle produce high pile upholstery fabrics, industrial furnishing fabrics and bags for vegetables. Tricot machines using bearded or compound needle produce lace, nets, and outerwear fabrics. Weft insertion with, for example, elastic yarns or fleeces can produce directionally orientated fabrics. Warp knitted fabrics are commonly used in linings for protective clothing and laminated with polyurethane foams to provide a strong flexible base for the foam.

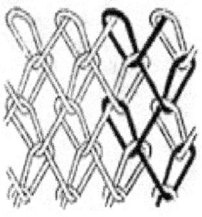

Figure 10 Warp knitted fabric construction

Warp knitting versus weft knitting

Warp knits, which generally have a flat, smooth surface (though they can also be made with a pile), have little or no vertical stretch and varying degrees of crosswise stretch. Produced in a large variety of weights in a wide range of fiber types, warp knits are run-resistant and don't ravel. With a few exceptions, weft knits have moderate to great amounts of crosswise stretch and some lengthwise stretch (some jerseys, however, have little or no crosswise or lengthwise stretch). On many weft knits, the edges may curl. As with warp knits, weft knits are made from many different fibers and come in many weights. If a stitch in a weft knit is broken, the fabric will tend to run, but a weft knit ravels only from the yarn end knitted last.

2.2.4 Multi-axial Fabrics

In recent years multi-axial fabrics have begun to find favour in the construction of composite components. The main fibres can be any of the structural fibres available in any combination (Figure 11). Multi-axials are nonwoven fabrics made with unidirectional fibre layers stacked in different orientations and held together by through-the-thickness stitching, knitting or a chemical binder.

Figure 11 Multi-axial Fabric

The proportion of yarn in any direction can be selected at will. In multi-axial fabrics, the fibre crimp associated with woven fabrics is avoided because the fibres lie on top of each other, rather than crossing over and under. This makes better use of the fibres inherent strength and creates a fabric that is more pliable than a woven fabric of similar weight. Super-heavyweight nonwovens are available (up to 200 oz/yd²) and can significantly reduce the number of plies required for a lay-up, making fabrication more cost-effective, especially for large industrial structures. High interest in non-crimp multi-axials has spurred considerable growth in this reinforcement category.

2.2.5 Braided Fabrics

Braid is a rope like thing, which is made by interweaving three or more strands, strips, or lengths, in a diagonally overlapping pattern. Braiding is a predominant manufacturing technique for developing textile reinforced hoses and ropes. Braiding is one of the major fabrication methods for composite reinforcement structures. Braiding is probably the simplest way of fabric formation. Diagonal interlacing of yarns forms a braided fabric. Each set of yarns moves in an opposite direction. Braided fabrics are continuously woven on the bias and have at least one axial yarn that is not crimped in the weaving process. The braid's strength comes from intertwining three or more yarns without twisting any two yarns around each other. This unique architecture offers, typically, greater strength-to-weight than woven. It also has natural conformability, which makes braid especially suited for production of sleeves and preforms because it readily accepts the shape of the part that it is reinforcing, thereby obviating the need for cutting, stitching or manipulation of fibre placement. Braids also are available in flat fabric form. These can be produced with a tri-axial architecture, with fibres oriented at 0°, +60°, -60° within one layer. This quasi-isotropic architecture within a single layer of braided fabric can eliminate problems associated with the layering of multiple 0°, +45°, -45° and 90° fabrics.

Furthermore, the propensity for delamination (layers of fibre separating) is reduced dramatically with quasi-isotropic braided fabric. Its 0°, +60°, -60° architecture gives the fabric the same mechanical properties in every direction, so the possibility for a mismatch in stiffness between layers is eliminated. In both sleeve and flat fabric form, the fibres are continuous and mechanically interlocked. Because all the fibres in the structure are involved in a loading event, the load is evenly distributed throughout the structure. Therefore, braid can absorb a great deal of energy as it fails. Braiding is the most widely used manufacturing technology in the development of rope and hose. Braid's impact resistance, damage tolerance and fatigue performance have attracted composite manufacturers in a variety of applications, ranging from hockey sticks to jet engine fan cases. Three-dimensional braiding is a relatively new topic, and mainly developed for industrial composite materials.

Braiding is done by intertwining of yarns in whatever direction suited to the manufacturer's purpose. Braiding can be classified as two and three-dimensional braiding. Two-dimensional braid structure can be circular or flat braid. They are formed by crossing a number of yarns diagonally so that each yarn passes alternately over and under one or more of the others. Two dimensional braids are produced through circular braiding machine and rotary machine (Figure 12).

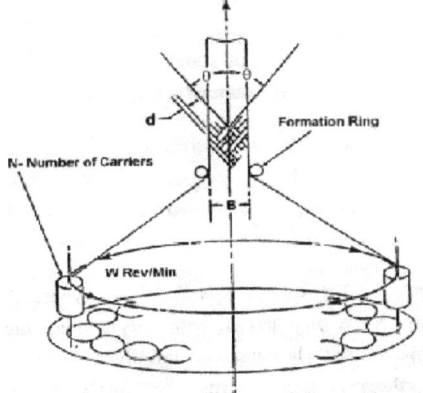

Figure 12 Brided structure formation in circular braiding machine

Three-dimensional braiding is relatively new and was developed mainly for composite structures. In it, a two dimensional array of interconnected 2-D circular braids is created on two basic types of machines the horn gear and cartesian machines. Braided fiber architecture resembles a hybrid of filament winding and woven material. Like filament winding, tubular braid features seamless fiber continuity from end to end of a part. Like woven material, braided fibers are mechanically interlocked (Figure 13). When functioning as a composite reinforcement, braid exhibits exceptional properties because it distributes loads very efficiently.

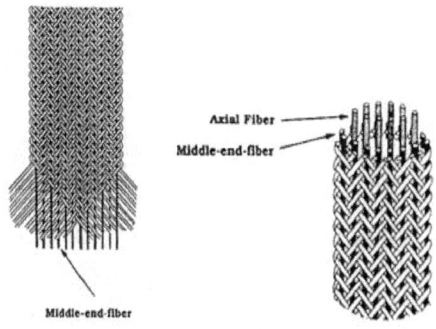

FIGURE 13 BRAIDED STRUCTURE

Types of braided structures

Braided textile structures are manufactured with mutual interwining of yarns in a tubular form. There are three typical braid structures: diamond, regular and hercules. Diamond structure is obtained when the yarns cross alternatively over and under the yarns of opposite direction. The repeat notation is 1/1. Regarding this way of notification, the regular braid structure has notation 2/2 and hercules 3/3. The braids are mostly produced in a regular structure. Generally braids are produced in a tubular form of biaxial yarns direction. By insertion of longitudinally oriented yarns (middle-end-fibre) into the structure the 3 axial braids is obtained. Moreover in the centre of the tubular braid, additional fibres called axial fibres can be inserted. When the number of braiding fibre bundles is the same, the tubular braid increases the fibre volume fraction more than the flat braid. The main feature of the braid is the angle of interwining that can vary between 10–80° and depends on: the yarn fineness, the type of the structure (biaxial or triaxial), cover factor (tightness of the structure) and the volume ratio of the longitudinal yarns. Since the braids have tubular form, they are often replaced with the filament winding structures. In this respect it has been proven that the braids can be competitive regarding the price. The braid is a flexible product and can be adjusted to various shapes. With the special device called mandrel the braids can be shaped into various forms directly on the machine at the manufacturing stage.

Braided composites, once used for such applications as drive shafts, propeller blades and sporting equipment, are becoming popular again in recent years partly due to the development of large computer controlled 2D and 3D braiders and partly due to the experience gained in using textile composites in the aerospace and automotive industries. Braiding has the potential to produce complex near-net shapes with fibre continuity at the edges and around holes and branches. However, unlike other quasi-laminar composites, the unit cell geometry of a braided composite is controlled by both the machine parameters and the component geometry.

Conventional 2D braided structure

Braided structures may be classified as (i) 2D braids produced on conventional maypole braiders, and (ii) 3D through-thickness braids produced on specialized machinery. 3D braiding was popular in the 1980s for aerospace applications. However, in recent years, composites industry has been taking a fresh look at 2D braids for developing affordable composite structures (Figure 14). For example, stitched 2D braided performs are being used for stiffeners and stringers in aircraft structures

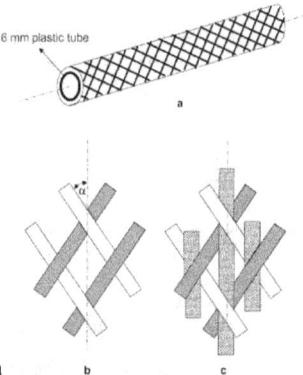

Figure 14 Triaxial braided structure for gears

Multi-axial differentially oriented structures (DOS)

Multi-axial differentially oriented structures (DOS) either using Karl Mayer's warp-knitted based method with variations in axially orientation of construction yarns or using LIBA's method of multiple weft-yarn stations give very interesting possibilities of producing technical textiles for a number of end-use applications. Karl Mayer's DOS structure incorporating thermoplastic yarns or split-films as matrix material has been used to produce high performance composites. This material is also suitable as substrate for coated products, and this technology allows incorporating nonwovens and other cellulose based material for introducing bulk in these structures. Because the inlaid yarns in DOS structures are placed straight without the any built-in crimp, the resultant stress distribution is an interesting factor in designing products for different applications where load-bearing aspect is important (Figure 15).

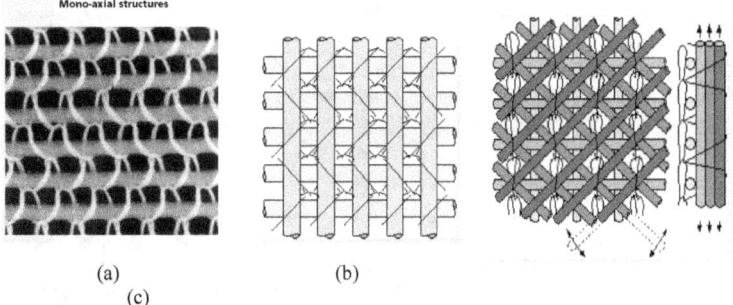

Figure 15 Braided structures (a) mono-axial, (b) bi-axial, and (c) multi-axial

Bi-axial structures
Bi-axial technical structures are formed as a combination of magazine weft insertion and a mislapping guide bar to obtain the lengthwise and widthwise reinforcement respectively.

Multi-axial structures
Multi-axial multi-ply structures are fabrics bonded by a loop system, consisting of one or several yarn layers stretched in parallel. Said yarn layers may have different orientation and different yarn densities of single ends. Multi-axial multi-ply fabrics are used to reinforce different matrices. The combination of multi-directional fibre layers and matrices has proved capable of absorbing and distributing extraordinarily high strain forces. The typical feature of warp-knitted multi-axial multi-ply structures for substrates is the interlacing of single ends in line with the stitch courses and the associated almost smooth processing of fed yarns. These products are multi-ply structures with angles of 30° up to 60° and/or 90°/0°. A multi-axial multi-ply fabric is demonstrating the two diagonal yarn sets in addition to the bi-axial yarn sets created by the mislapping guide bar (warp or ST-yarn) and the magazine weft insertion system (weft).

Multi-axial structures for fibre reinforced plastics
The typical feature of warp-knitted multi-axial multi-ply structures for fibre reinforced plastics is the stitching-through principle, ensuring a uniform distribution of yarn ends and preventing gap formation. In addition, fibrous webs, films, foams and other materials can be incorporated. Angle positions of -45° through 90° up to +45° and 0° are generally used, being infinitely variable. Due to the non crimped and parallel yarn sheets, such as multi-axial multi-ply structures are particularly suitable to reinforce plastics to form fibre reinforced plastics.

The special characteristics of such composites are:
- low specific weight
- adjustable stiffness between extremely stiff and extremely stretchable
- resistance to corrosion and chemicals
- highest mechanical load resistance

Applications:
- Rotor blades for wind power stations.
- Moulded parts for automotive, aircraft, and ship building.
- Equipment for sports and leisure-time activities e.g. skis, snowboards, surfboards, sports boats

2.2.6 Nonwoven Structures
Nonwoven technology is a newer one regarding other technologies of fibre forming assemblies. A broader definition is that nonwoven is a flexible structure manufactured by bonding or interlocking of fibres, or both, accomplished by mechanical, chemical, thermal or solvent means and combination of thereof. Nonwoven products make up more than 35 percent of industrial textile processing and more than 90 percent of the nonwoven fabric is used for industrial textiles. The versatile and innovative nature

of these nonwoven fabrics makes them an increasingly important raw material for industrial textiles. Nonwoven technology has made a breakthrough in the areas of isotropy, uniformity, hand feel, and thickness of the products. Nonwoven filter media are produced by forming a mat of fibres. The fiber diameter, orientation, packing density and web weight all determine the filter media properties. The smaller the nonwoven web pore size the finer the filtration efficiency. Nonwoven fabrics dominate nearly more than 90% of the filtration market, the reason attributed are variability and economical manufacture they can easily adapted to all kinds of filtration jobs. The nonwovens generally are produced with polyolefin, polyester or nylon polymers and fibres. Melt blown media are one of the most versatile nonwovens for liquid filtration and the only one that can achieve submicron filtration. Melt blown media have nominal ratings from 1 μm to 50 μm and when calendared or laminated into composites can have sub micron and absolute ratings. Wet laid media are generally produced on a paper machine and with cellulose, polymeric or glass fibres. High efficiency wet laid media have equivalent efficiency as microfiltration membranes but with significantly higher dirt holding capacity or life.

Fabrics made by nonwovens technology can be made up to five times more durable than conventional textile fabrics of the same weight. They can be designed to be extremely abrasion and heat resistant. Some fabrics can withstand extremely high temperatures for example; mechanical bonded glass fibres can be used at operating temperatures up to 1000°F and silica materials can be used up to 2000°F. For acoustic insulation, nonwoven webs are used which weigh 50% less than any comparable material and provide the same or higher absorption values. Nonwoven webs have high barrier properties; they can filter almost anything ranging from macro to nano scale particle sizes. The nonwovens have a broad range of performance: from light materials for wadding and insulation where the fibre volume ratio equals only 2–3 %, to compact fabrics for reinforcement where the fibre volume ratio comes up to 80 %. Multilayer combinations of woven and nonwoven fiber web can serve as noise absorption elements in a wide range of applications including acoustic ceilings, noise reducing quilts and noise proof barriers.

Nonwovens are a class of fabric that are produced directly from fibres, and in some cases directly from polymers, thereby obviating a number of intermediate processes such as spinning, winding, warping, weaving/knitting. Hence nonwovens can be produced inexpensively for both single use and durable applications. Nonwovens are produced in two distinct steps:

1. Web formation: arrangement of fibres into a 2D sheets, and
2. Consolidation: bonding the fibres together to create a nonwoven fabric.

Web formation methods
Web formation may be classified into dry laid, spun laid and wet laid processes.

Dry laid process
Dry textile fibres are carded, using a carding machine similar to the once used in the spinning industry, to arrange the fibres in a 2D sheet with fibre orientations

predominantly in the machine direction. The web is subsequently folded using a cross-lapping machine to increase the web thickness and to achieve transverses fibre orientation. In some cases, conventional carded and cross-lapped webs are combined to produce a web with bi-directional fibre orientation. Alternatively, an aerodynamic system is used for creating a web random fibre orientation.

Spun laid process

This is a method of producing fabrics directly from polymer chips, hence eliminating the entire textile supply chain. Fibres are extruded from a spinneret similar to conventional melt spinning process. These fibres are attenuated (stretched) using high-velocity air streams before depositing on a conveyer in a random manner. The spun laid process is the most commonly used method for producing both disposable and durable nonwovens for protective application. There are other related systems such as flash spinning, melt blowing and electro-spinning. Flash spinning involves extrusion of a polymer film dissolved in a solvent; subsequent evaporation of the solvent and mechanical stretching of the film results in a network of very fine fibres. These fibres are subsequently bonded to create a smooth, micro porous textile structure used for protective applications. The melt-blowing process produces micro fibres by attenuating the polymer jet, coming out of the spinneret, using high-velocity air jet. Since the polymer is stretched in the molten state, extremely fine fibres can be produced. Because of the lack of molecular orientation, melt-blown fibres are weak and hence are generally used in conjunction with other type of non-weaves. For example, a composite nonwoven consisting of melt-blown layer and a spun-bond layer is becoming popular for medical protective applications

Wet laid process

Developed from the traditional paper making process, relatively short textile and wood fibres are dispersed in large quantities of water before depositing on as inclined wire mesh. These materials find application in hospital drapes and filters.

Consolidation process

Fibrous webs can be consolidated using a number of techniques depending on the area density and the desired properties. They can be classified into mechanical, chemical, thermal and stitch bonding processes.

Mechanical bonding

Needle punching and hydro-entanglement are two complementary mechanical processes. Relatively thick webs (150 to 1000 g/m2) are felted with the aid of oscillating barbed needles. The hydro-entanglement process uses high velocity water jets to consolidate relatively thin webs (< 140 g/m2). The resulting spun laced fabrics are highly drapable and hence popular for medical protective clothing.

Chemical bonding

Fibres are bonded with a suitable adhesive and subsequently cured under heat. Saturation bonding is seldom used for protective applications, as this process results in a relatively stiff non-porous material. Spray and print bonding instead of saturation bonding improves the flexibility and permeability.

Thermal bonding.
Relatively thin webs are passed through a heated calender, resulting in partial melting and bonding of fibres. Thermal bonding is a high-speed process and hence commonly used in conjunction with spun laids.

Stitch bonding
Cross-laid webs are stitched together with a relatively large number of needles across the width. Alternatively, stitch bonding is also used to bond a series of non-interlaced thread systems.

For making staple nonwovens, fibers are first spun and cut into staple form and then compressed into bales. These bales are then opened and scattered on a conveyor belt, and the fibers are spread in a uniform web by a wet laid process or by carding. These nonwovens are either bonded thermally or by using resin. The spun laid nonwovens are made in one continuous process (Figure 16). Fibers are spun and then directly dispersed into a web by deflectors or with air streams. Melt blown nonwovens have extremely fine fiber diameters but are not strong fabrics. Spun laid is also bonded either thermally or by using resin. Both staple and spun bonded non-wovens would have no mechanical resistance without the bonding step. Nonwoven fabric has gradually gained importance in various industrial applications along with medicine, personal care, hygiene and household uses. They are used in interlinings and apparel, carpet backing and underlay, needle punched felt for backing of PVC floor covering, home furnishing and household products, medical, sanitary and surgical applications, book cloths, Industrial wiping cloths, filtration, shoe linings, automotive applications, laundry & carry bags in hospitality industry etc.

FIGURE 16 NON WOVEN FABRIC MADE OUT OF POLYPROPYLENE

2.2.7 Membranes

Microfiltration (MF) membranes are frequently used as a pretreatment for reverse osmosis systems. MF membranes are available in pleated cartridge or hollow fiber formats at a range of efficiencies. They can be run in normal or cross flow mode and many hollow fiber formats are back flushable. The development of back flushing hollow fibres has made membrane based water filtration an economic process. Microfiltration membranes are produced by a wide variety of processes. The most common process is to dissolve a polymer in a solvent and produce a liquid film on a support material. The solvent is removed by evaporation or dilution in a non solvent and the polymer is precipitated forming a porous structure. The concentration of polymer in the solution and the rate of precipitation determine the degree of porosity and the pore size distribution. Microfiltration membranes are produced with efficiencies from 0.05 μm to 5 μm.

Industrial Textile Products Manufacturing Processes

There are several ways for manufacturing flexibility in production/ manufacturing of industrial products which are shown in Figure 17.

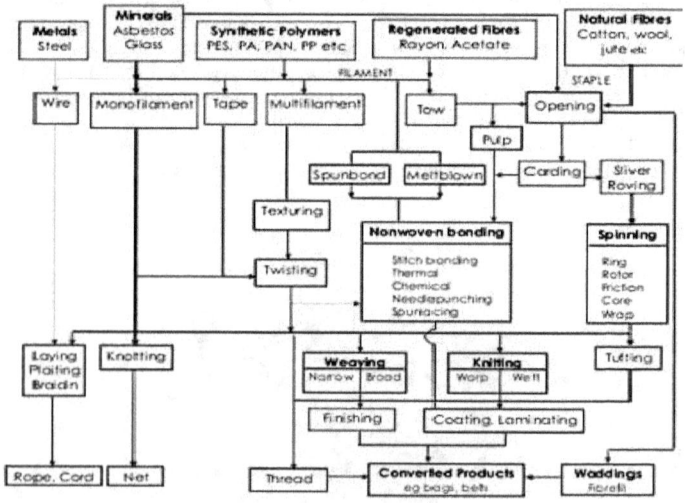

Figure 17 Industrial Textile Products Manufacturing Processes

The products of industrial textile, since the general definition of textile as an assembly of textile fibres into useful product is very broad and can be spread onto another segments of flexible industrial engineering. It utilize high-performance fibres, where the bulk of the products use conventional polymer materials such as polyethylene (PET), polypropylene (PP), polyamide (PA), viscose, cotton, jute and even glass is surprising. However, the properties and the structure of conventional fibres have been substantially modified compared to the ones used in everyday textile. The technology of industrial textile is not always confined to the products themselves or to the production technology but it incorporates "know-how" for the application of these products for a broad range of end users such as: aviation, industry, medicine, defense, security, transport, construction, agriculture, etc. The fact that leads to confusion is that manufacturing of many products (for example, manufacturing of metal wires covered with braids or fabrics) employs pure textile techniques.

REFERENCES
[1] A. F. Turbak, T. I. Vigo, in High-Tech Fibrous Materials, T. I. Vigo and A. F. Turbak, eds., American Chemical Society, Washington, DC, 1991, pp. 1–15.
[2] A R Horrocks, and S C Anand: Handbook of Technical Textiles, Woodhead Publishing, Cambridge, 2000.
[3] R. Caleske, Focus on design: Carbon fibre epoxy violin captures critical acclaim, High-Performance Composites, 12, 2, 14 (2004).
[4] T. Matsuo, Overiew and trend of technical textile technologies in Japan, Nonwovens & Industrial Textiles, 48, 1, 46–48 (2001).
[5] B. Piller, Integral knitted fabrics with increased moisture transfer, Knitting Technique, 9, 358–364 (1987).
[6] C. Byrne, in High Technology Fibers, A. Horrocks, S. C. Anand (eds.), Woodhead Publishing Ltd., Cambridge 2000, pp. 6–9.
[7] Demboski, G. Bogoeva-Gaceva, Textile structures for technical textiles, Bull. Chem. Technol. Macedonia, 24, 1, 67–75 (2005).
[8] A Ormerod, and W S Sondhelm: Weaving -- Technology and Operations, Textile Institute, Manchester, 1999.
[9] D J Spencer, Knitting Technology, Pergamon Press, Oxford, UK, 1986.

www.ingramcontent.com/pod-product-compliance
Lightning Source LLC
Chambersburg PA
CBHW072049190526
45165CB00019B/2246